THE KIDS' FORT BOOK

BY DAVID STILES

To my daughter Lief-Anne
who was the inspiration
for all of these forts.

THE KIDS' FORT BOOK is an original publication of Avon Books. This work has never before appeared in book form.

7th grade reading level has been determined by using the Fry Readability Scale.

AVON BOOKS
A division of
The Hearst Corporation
959 Eighth Avenue
New York, New York 10019

Published by arrangement with the author
Library of Congress Catalog Card Number: 81-52389
ISBN: 0-380-79277-x

Cover photograph by David Stiles

First Camelot Printing, April, 1982

Library of Congress Cataloging in Publication Data
Stiles, David R.
The kids' fort book.

(An Avon Camelot book)
Summary: Includes directions for building indoor and outdoor forts from blankets, boxes, leaves, sand, snow, and logs, as well as instructions for making model forts and toy weapons.
1. Fortification—Juvenile literature. [1. Fortification. 2. Fortification—Models. 3. Models and model-making. 4. Toy making. 5. Handicraft] I. Title.
UG401.S84 745.592'82 81-52389
ISBN 0-380-79277-X AACR2

Printed in the U. S. A.

DON 10 9 8 7 6 5 4 3

CONTENTS

INTRODUCTION

THE MEANING OF THE WORD FORT HAS CHANGED HISTORICALLY AS CIVILIZATION HAS CHANGED. ALL SOCIETIES HAVE DEVELOPED WAYS OF DISPLAYING THEIR AUTHORITY, AND OF PROCTECTING THEIR COMMUNITIES. THEY HAVE DIFFERED ONLY IN THEIR INVENTIVENESS.

THE MEDIEVAL FORT, OR NORMAN CASTLE, WAS BUILT FOR PROTECTION AGAINST INVASION, AND BECAME A FORTIFIED RESIDENCE. IT WAS A FAMOUS STYLE THAT INCLUDED THE MOAT AND DRAWBRIDGE. THESE CASTLES WERE OFTEN BUILT ON TOP OF THE RUBBLE FROM A DEVASTATED VILLAGE, ENABLING THEM TO HAVE AN UNINTERRUPTED VIEW OF THE SURRONDING COUNTRY-SIDE AND ANY ENCROACHING ENEMY.

MEDIEVAL FORT

THE MUCH IMITATED STYLE OF THE NORMAN CASTLE WAS INTRODUCED IN 1066 WITH THE INVASION OF THE NORMANS IN BRITAIN. THEIR USE OF STONE AS A BUILDING MATERIAL REPLACED THE ANGLO-SAXON WOOD STOCKADE.

STOCKADE FORT

IN THE UNITED STATES, FORTS WERE BUILT ALONG TRAILS TO PROTECT TRAVELERS, PROVIDE A PLACE OF REFUGE IN CASE OF INDIAN ATTACK, AND TO ESTABLISH TRADING POSTS. FRONTIERSMEN BUILT FORTS OUT OF THE ABUNDANT TIMBER IN THE REGION. OFTEN THESE FORTS BLOOMED INTO THRIVING COMMUNITIES AS MORE AND MORE SETTLERS TRAVELED ACROSS THE COUNTRY.

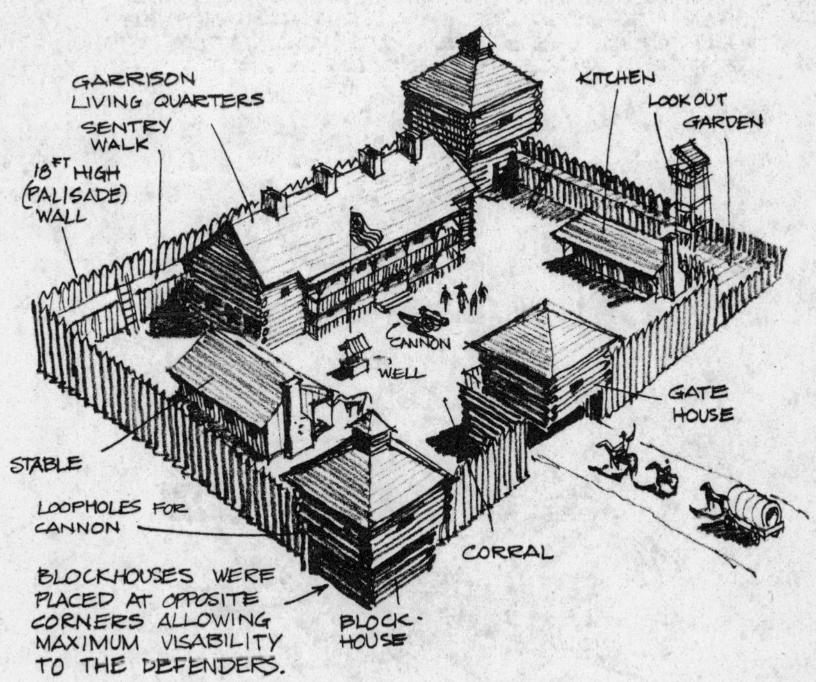

FORT OF THE FUTURE

FORTS OF THE FUTURE ARE OPEN TO YOUR IMAGINATION. THEY MAY BE SHINY SKYSCRAPERS WITH STAINLESS STEEL BASTIONS AND COMPUTERIZED LAZER GUN TURRETS.

IMAGINATION WILL PLAY A GREAT PART IN THE VISION AND CREATION OF YOUR OWN FORT. A SIMPLE "BLANKET OVER TWO CHAIRS" FORT CAN BECOME A MIGHTY MAGICIAN'S UNDERGROUND CAVERN, AND A SPIRALED SAND CASTLE CAN BECOME THE ROMANTIC FORTRESS OF A PRINCESS AND HER ENTOURAGE. SO USE YOUR IMAGINATION AND ENTER A WHOLE NEW WORLD....

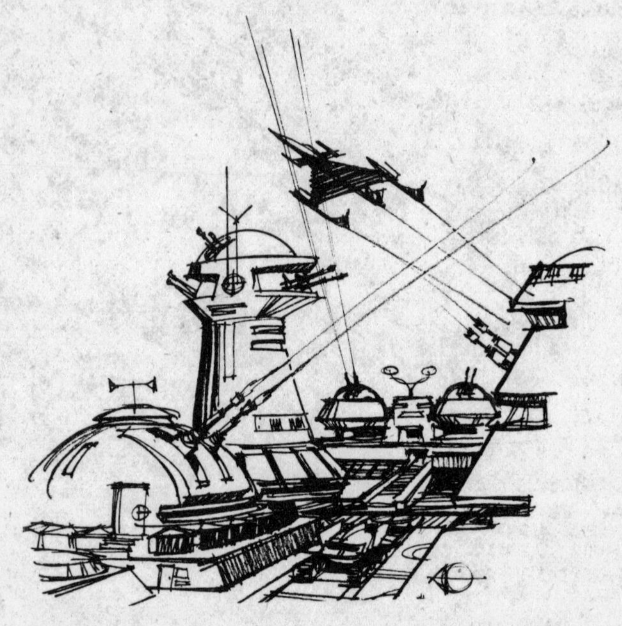

KIDS' FORTS

A FORT IS A VERY SPECIAL PLACE WHERE YOU CAN RELAX AND FEEL SAFE. IT IS A PLACE TO BE BY YOURSELF IN YOUR OWN WORLD OR SHARE WITH A VERY GOOD FRIEND. IT IS A PLACE THAT IS JUST YOUR SIZE - MADE JUST FOR YOU.

ONCE YOU ARE COMFORTABLE IN YOUR FORT YOUR IMAGINATION CAN CHANGE IT INTO ANYTHING FROM A GIANT'S DUNGEON, TO A SECRET CAVE BENEATH THE EARTH, OR A SPACE SHIP TAKING YOU ON A FANTASTIC VOYAGE THROUGH THE STARS.

KIDS: CONSULT YOUR PARENTS BEFORE BUILDING ANY OF THESE FORTS, AND BEFORE USING ANY TOOLS WHICH ARE UNFAMILIAR TO YOU.

HOUSEHOLD FORTS

IT DOES NOT TAKE MUCH TO MAKE A FORT. SOMETIMES ALL YOU NEED IS A BLANKET. HERE ARE SOME WAYS YOU CAN MAKE A FORT BY JUST USING A BLANKET AND PILLOWS.

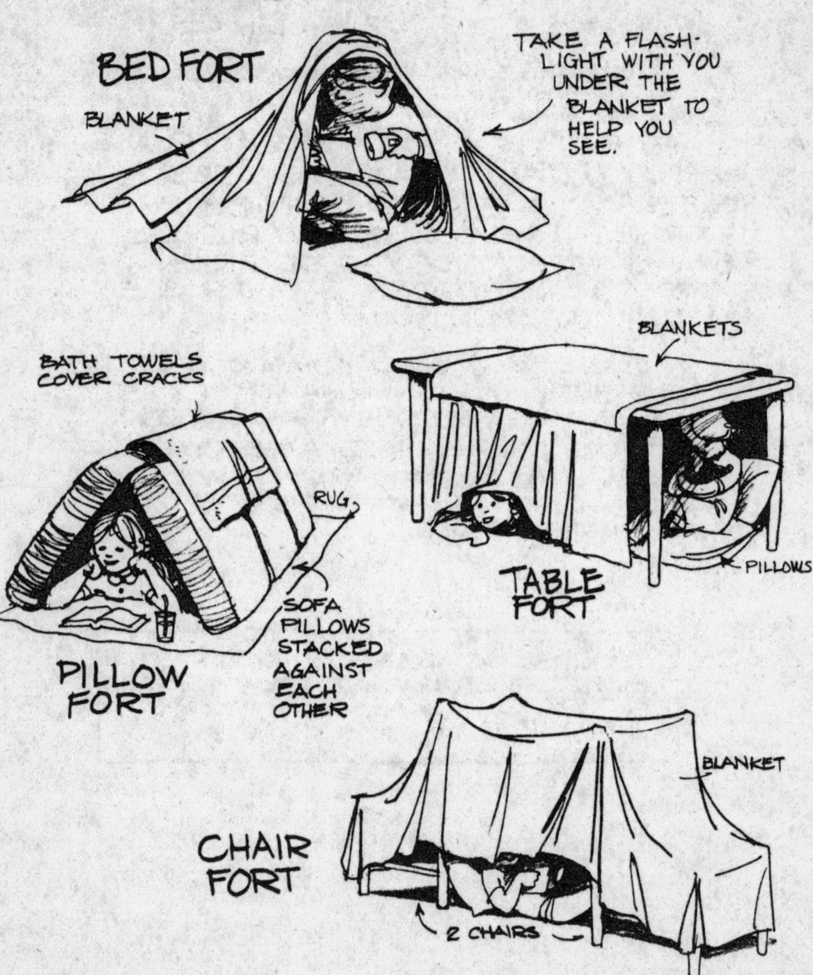

REFRIGERATOR BOX FORT

NOTE:
USE ENAMEL PAINT IF FORT IS TO BE USED OUTSIDE.

FIND AN OLD REFRIGERATOR CARTON AND ASK MOM OR DAD TO HELP YOU CARRY IT INTO THE HOUSE

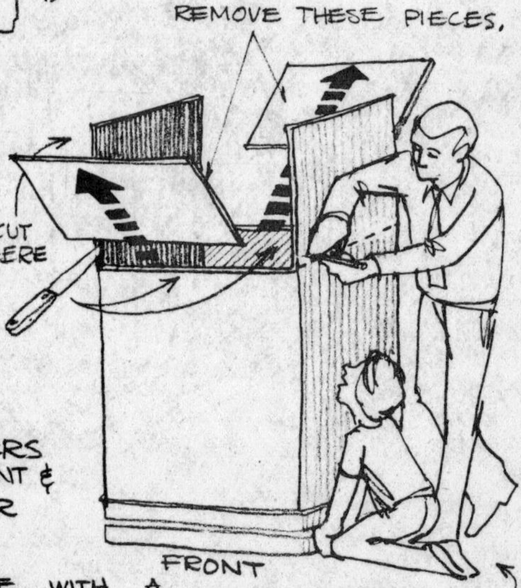

DISCARDED CARDBOARD REFRIGERATOR CARTON

DAD

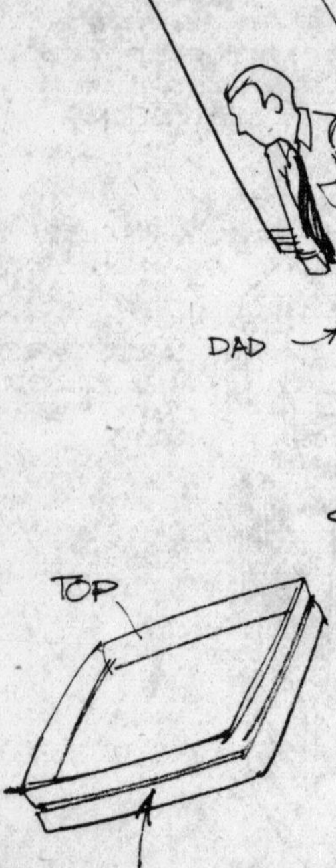

REMOVE THESE PIECES.

TOP

CUT HERE

REMOVE THE TOP.
ASK MOM OR DAD TO CUT DOWN THE CORNERS AND ACROSS THE FRONT & REAR WITH A BUTCHER KNIFE.

FRONT

THEN MARK A LINE WITH A PENCIL FROM FRONT TO REAR AS SHOWN HERE.

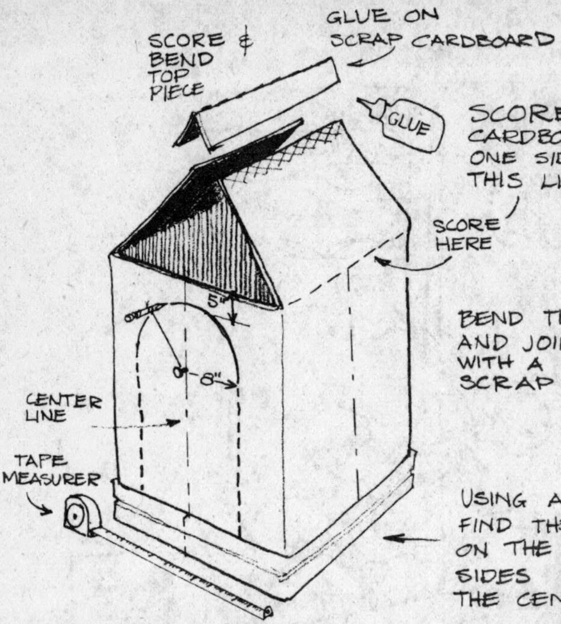

GLUE ON SCRAP CARDBOARD

SCORE & BEND TOP PIECE

GLUE

SCORE - OR CUT THE CARDBOARD SLIGHTLY ON ONE SIDE ONLY, ALONG THIS LINE.

SCORE HERE

5"

8"

CENTER LINE

BEND THE TOP SIDES IN AND JOIN THEM TOGETHER WITH A BENT PIECE OF SCRAP CARDBOARD.

TAPE MEASURER

USING A TAPE MEASURER FIND THE HALFWAY MARK ON THE FRONT AND TWO SIDES DRAW A LINE UP THE CENTER OF EACH SIDE.

FROM THE CENTER LINE DRAW A 16" HALF CIRCLE USING A COMPASS MADE FROM A STRING AND A PENCIL.

CUT THE HALF CIRCLES WITH A KNIFE. ALSO CUT THE CENTER LINE AND THE BOTTOM.

SCORE THE HINGE SIDE OF THE DOORS AND OPEN THEM UP.

FOLLOW THE SAME PROCEEDURE FOR THE SIDE WINDOWS.

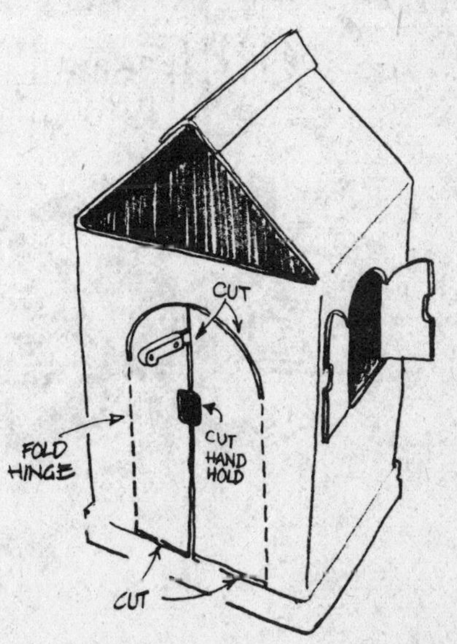

CARD-BOARD BOX FORT

THIS FORT IS MADE ENTIRELY OF CARDBOARD BOXES FOUND AT SUPER-MARKETS AND LIQUOR STORES. SOME CARDBOARD BOXES ARE SO STRONG THAT THEY CAN SUPPORT A LARGE MAN. THESE ARE THE KINDS THAT YOU WILL FIND AT LIQUOR STORES. THEY HAVE CARDBOARD DIVIDERS IN THEM.

BE SURE TO KEEP THE DIVIDERS WHEN YOU TAKE THE BOXES, AND PLACE THEM VERTICALLY WHEN YOU BUILD THE FORT.

OTHER BOXES, LIKE PAPER TOWEL BOXES, ARE WEAK AND SHOULD BE STRENGTHENED BY INSERTING DIVIDERS FROM LEFTOVER BOXES.

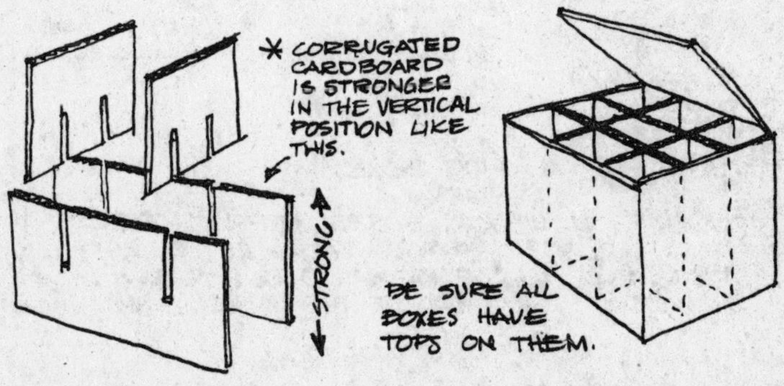

TOOLS

ALL YOU WILL NEED TO MAKE THE FORT ARE:

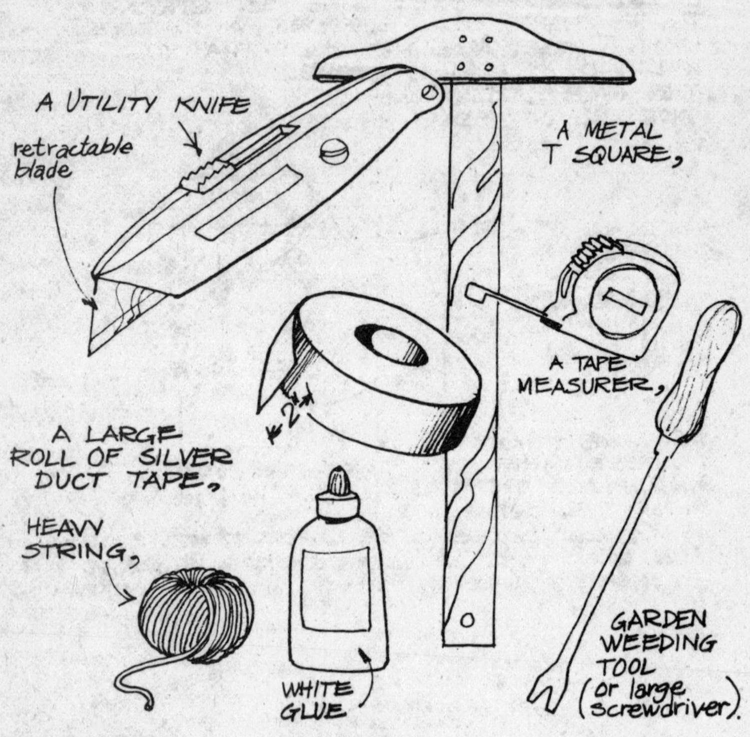

BOXES:

1ST LAYER - 12 LIQUOR BOXES 13" HIGH

2nd LAYER - 16 LIQUOR BOXES APPROXIMATELY THE SAME HEIGHT

ENTRANCE - 3 LARGE PAPER TOWEL BOXES - TWO OF THEM MUST BE THE SAME HEIGHT WHEN LAID ON THEIR SIDES

NOTE: ALL BOXES MUST HAVE LIDS.
TAKE SOME EXTRA BOXES TO MAKE DIVIDERS.

BEGIN BY MEASURING ALL THE LIQUOR BOXES AND FIND 12 THAT ARE THE SAME HEIGHT.

ARRANGE THE BOXES IN AN OPEN CIRCLE WITH THEIR LIDS POINTING OUT FROM THE CENTER OF THE CIRCLE.

TAKE TWO PAPER TOWEL BOXES AND REINFORCE THEM WITH CARDBOARD DIVIDERS. SET THEM IN FRONT OF THE OPEN CIRCLE TO FORM AN ENTRANCE. AND PLACE A THIRD LARGE BOX ON TOP.

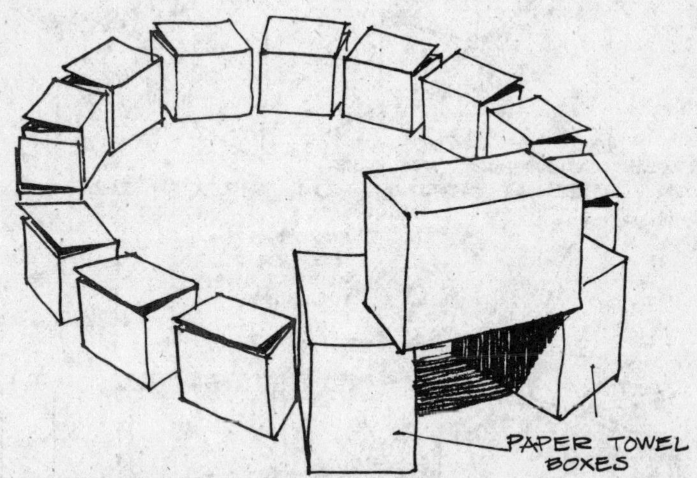

ENTRANCE

THE ENTRANCE BOXES ARE JOINED BY LACING THEM TOGETHER WITH HEAVY STRING. USING A WEEDING TOOL OR LARGE SCREWDRIVER, POKE HOLES THROUGH THE BOTTOM AND TOP BOXES. LACE A SHORT LOOP OF STRING THROUGH THE HOLES AND TIE THE BOXES TOGETHER.

(SEE ILLUSTRATION ON NEXT PAGE.)

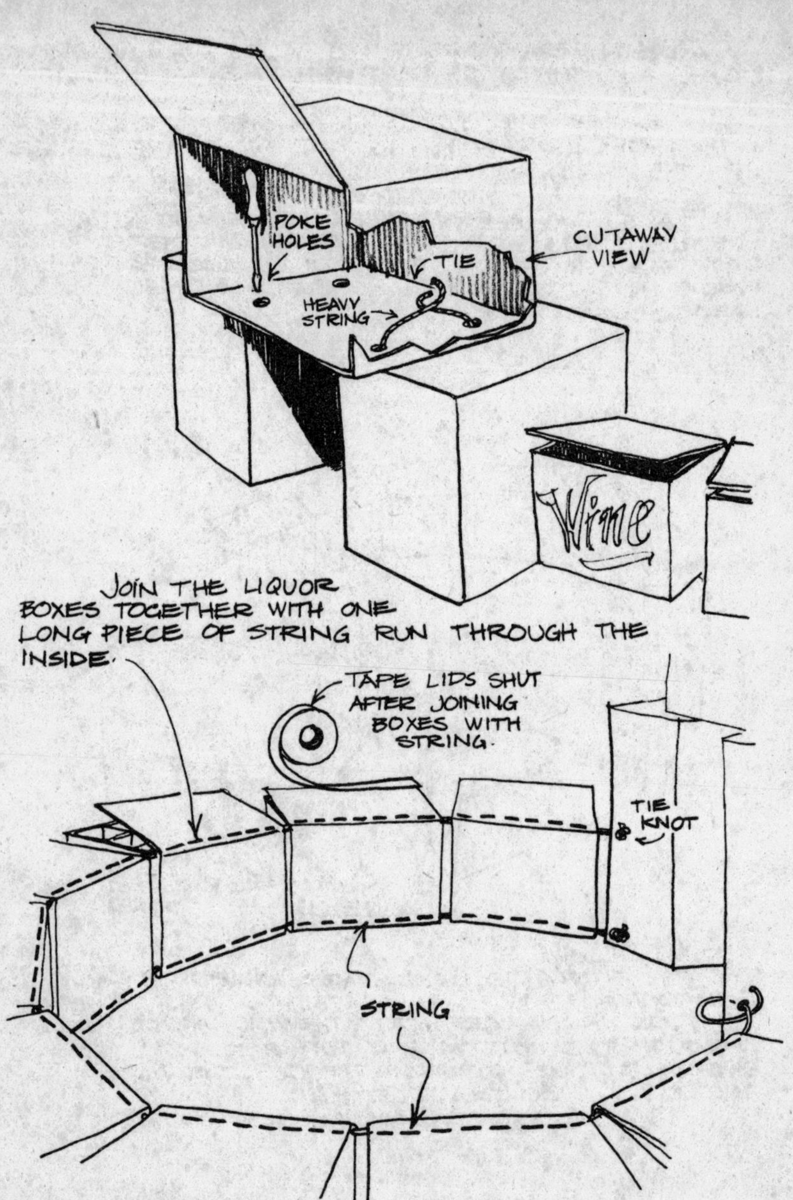
POKE HOLES
TIE
CUTAWAY VIEW
HEAVY STRING
Wine
JOIN THE LIQUOR BOXES TOGETHER WITH ONE LONG PIECE OF STRING RUN THROUGH THE INSIDE.
TAPE LIDS SHUT AFTER JOINING BOXES WITH STRING.
TIE KNOT
STRING

ADD ANOTHER LAYER OF BOXES BY GLUING & TAPING THEM ONTO THE FIRST LAYER.

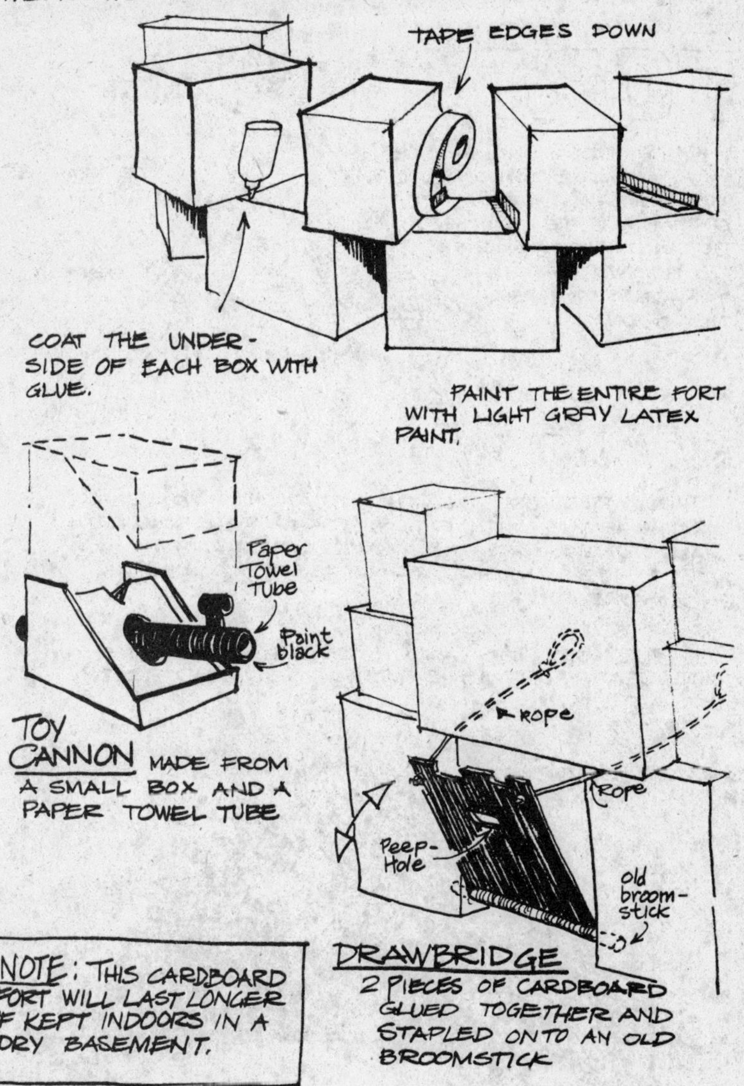

NOTE: THIS CARDBOARD FORT WILL LAST LONGER IF KEPT INDOORS IN A DRY BASEMENT.

LEAF FORT

FIND THREE LONG BRANCHES AND TIE THE ENDS TOGETHER WITH ROPE OR WIRE. SPREAD OUT THE BRANCHES. TIE ANOTHER BRANCH ACROSS TWO OF THE BRANCHES TO FRAME THE DOOR.

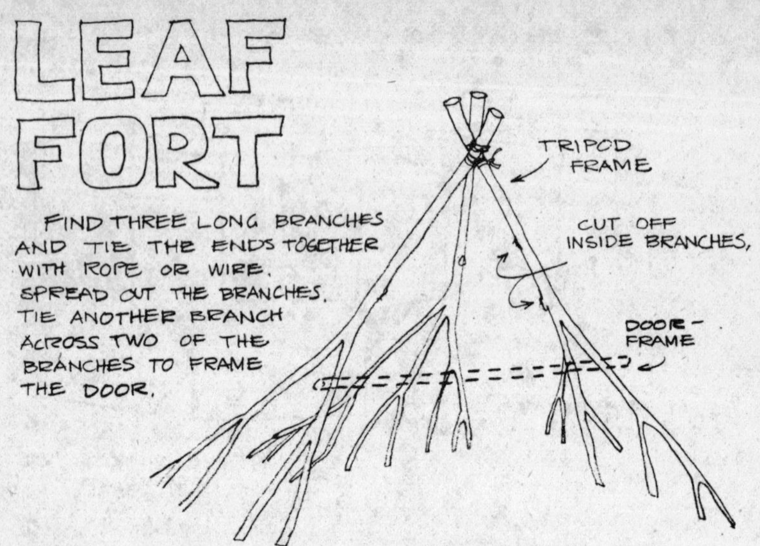

THEN PILE UP ALL THE BRANCHES YOU CAN FIND, STARTING WITH BIG ONES AND FINISHING WITH SMALL ONES. TRY TO WEAVE SMALLER BRANCHES IN AND OUT OF THE LARGER BRANCHES TO FORM AN ENCLOSURE.

PILE UP LOTS AND LOTS OF LEAVES. NOW YOUR LEAF FORT IS CAMOUFLAGED AND READY FOR ACTION.

SAND FORT

SAND FORTS CAN BE MADE QUITE EASILY IF YOU REMEMBER TO BRING THE PROPER EQUIPMENT WITH YOU TO THE BEACH. SOME OF THE THINGS YOU MIGHT NEED ARE:

YOU CAN HAVE A MOAT (WATER) AROUND YOUR SAND FORT IF YOU PLAN IN ADVANCE.

1. USE A TIDE TABLE FIND WHEN THE NEXT HIGH TIDE WILL BE. (PLAN TO HAVE YOUR FORT FINISHED BY THAT TIME.)
2. TRY TO FIGURE HOW HIGH THE TIDE WATER LEVEL WILL COME UP.
3. DIG AN 8 FOOT DIAMETER MOAT DEEP ENOUGH SO THAT AS THE TIDE COMES UP, THE WATER LEVEL (BENEATH THE TOP SURFACE OF THE SAND) WILL FILL YOUR MOAT AUTOMATICALLY. WHEN YOU ARE DIGGING OUT THE MOAT, PILE THE SAND ON THE INSIDE OF THE CIRCLE TO FORM THE WALLS OF THE FORT.
4. MAKE A ROOF WITH BOARDS AND COVER WITH SAND.

MOAT

EXPECTED HIGH TIDE WATER LEVEL

WATER LEVEL

SNOW-BALL FORT

SNOWBALL FORTS ARE MADE BY ROLLING BIG BALLS OF SNOW TO ONE LOCATION AND PILING THEM ON TOP OF EACH OTHER TO FORM A FORT. BOARDS PLACED ACROSS THE TOP FORM THE ROOF, AND A SECOND WALL IN FRONT OF THE MAIN STRUCTURE FORMS A PROTECTION-BARRIER FROM ATTACK. BE SURE TO PROVIDE SMALL WINDOWS AROUND THE FORT SO YOU CAN OBSERVE POTENTIAL INVADERS. BUILD A SMALL LEDGE ON THE INSIDE OF THE WALL TO HOLD SNOW-BALLS IN READINESS AND HAVE AN EXTRA SUPPLY IN AN AMMUNITION SLED MADE OUT OF AN OLD WOODEN FRUIT BOX.

CITY SNOW FORT

LUCKY ARE YOU IF YOU LIVE IN THE CITY DURING A HEAVY SNOWSTORM. WAIT FOR THE SNOW TRUCKS TO FINISH CLEARING THE STREETS AND LOOK FOR A LARGE PILE WHERE THEY HAVE DEPOSITED THEIR LOAD OF SNOW. PLACE THREE OR FOUR BOARDS ACROSS THE TOP, SHOVEL OUT THE SNOW FROM UNDERNEATH AND PILE IT ON TOP. IN 15 MINUTES YOU WILL HAVE A SAFE SNOW FORT.

SNOW BLOCK FORT

THE SNOW BLOCK FORT IS MADE BY FIRST MAKING A BOX THAT IS USED AS A MOLD FOR BUILDING THE FORT. THE BOX IS MADE OF ½" PLYWOOD, NAILED TOGETHER WITH 1½" FINISHING NAILS AND VARNISHED. TO BUILD THE FORT, PACK THE BOX WITH SNOW AND DRAG IT TO THE FORT SITE. TURN THE BOX UPSIDE DOWN AND A NICE NEAT BLOCK OF SNOW WILL DROP OUT. STACK UP THE BLOCKS AND MAKE A CIRCULAR FORT.

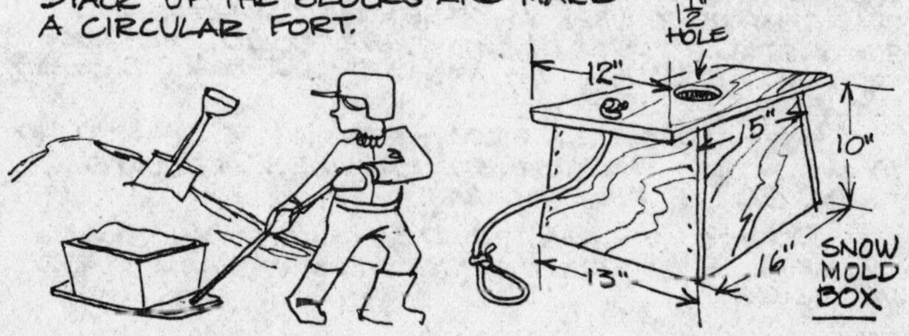

NOTE: THE FOLLOWING THREE FORTS INVOLVE SOME COMPLICATED INSTRUCTIONS AND CARPENTRY. ASK YOUR PARENTS FOR HELP.

BUNKER FORT

STRUCTURES LIKE THIS HAVE BEEN BUILT FOR CENTURIES BY SHEEPHERDERS, WINE GROWERS, AND FARMERS ALL OVER THE WORLD. DURING WARTIME THEY WERE USED AS OBSERVATION POSTS AND GUN INSTALLATIONS DUE TO THEIR SOLID CONSTRUCTION AND BECAUSE THEY ARE DIFFICULT FOR THE ENEMY TO OBSERVE.

BUILT ALMOST ENTIRELY FROM MATERIALS FOUND IN THE WOODS, THE BUNKER PROVIDES ADEQUATE SHELTER AT MINIMUM COST.

DURING THE WINTER THE BUNKER CAN BE CLOSED UP IN FRONT TO PROVIDE A WARM PLACE TO SLEEP.

FIND A SECLUDED SPOT WITH A TERRIFIC VIEW, PREFERABLY ON A HILLTOP. BEGIN BY DIGGING OUT THE SIDE OF THE HILL AND PILING THE DIRT IN FRONT.

DIG OUT AN AREA 5 FT. SQUARE FOR THE BUNKER AND PROVIDE A 4 FT. FLAT AREA IN FRONT.

REMOVE EARTH HERE

... AND ADD IT HERE.

4'
FLAT AREA

5'
BUNKER AREA

5'

FIND FOUR HEAVY FORKED BRANCHES IN THE WOODS AND CUT THEM 5 FT. LONG. DIG FOUR 12" DEEP HOLES IN A SQUARE 5 FT. APART AND BURY THE POLES IN THE GROUND.

- LAY A CROSSBEAM ACROSS EACH PAIR OF POSTS.
- NEXT BUILD A WALL BEHIND THE BACK POSTS WITH BOARDS OR LOGS TO FORM A REAR WALL.
- BUILD THE SIDE WALLS IN THE SAME MANNER. WHEN YOU HAVE BUILT THE WALLS 12" HIGH, STOP AND LAY SOME FLOORBOARDS ACROSS- THEN CONTINUE THE WALL TO THE TOP.

NEXT FIND A FAT LOG FOR THE RIDGEPOLE. LAY IT FRONT TO REAR ON THE TWO CROSS LOGS- THEN COVER THE ROOF WITH BOARDS.

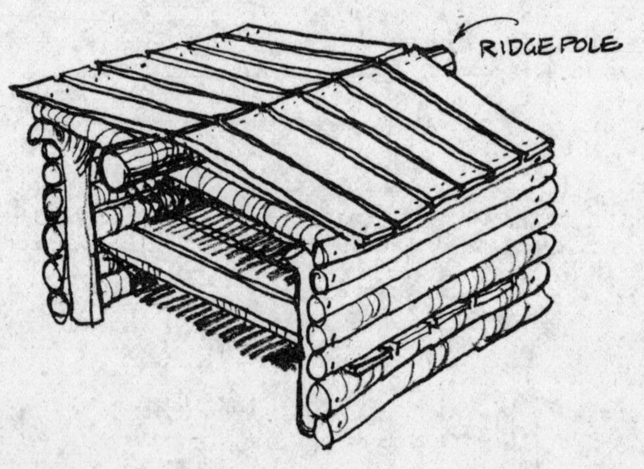

STARTING FROM THE BOTTOM, STAPLE TAR PAPER OVER THE OUTSIDE WALLS AND ROOF.

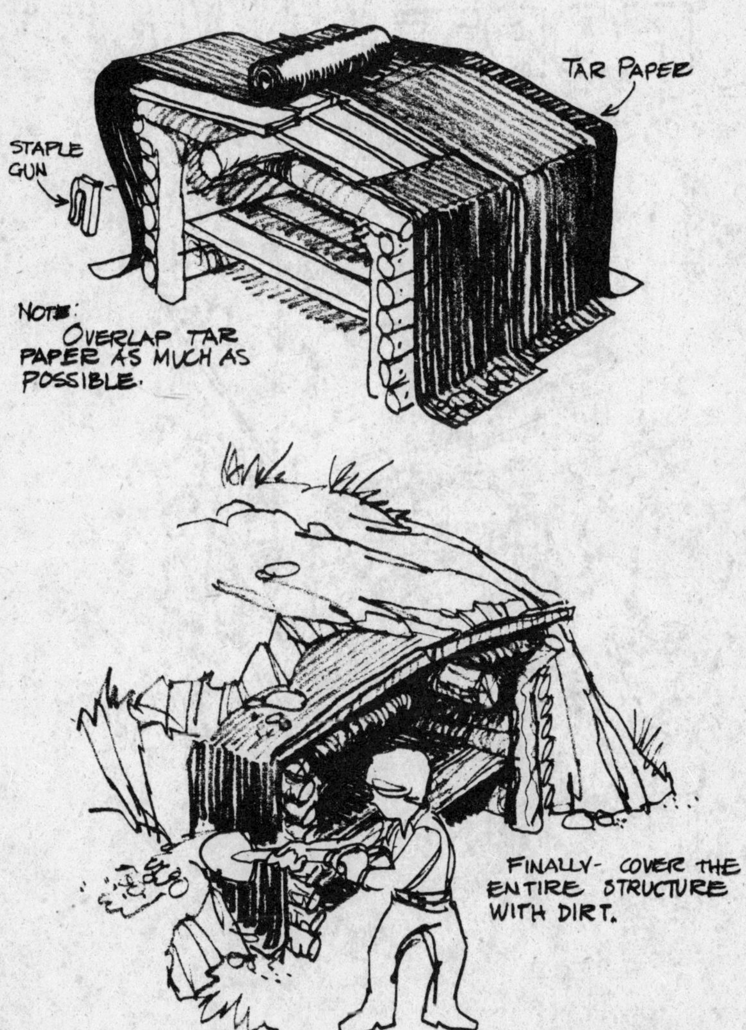

TREE
FORT
CLUB 3

FIND THREE TREES 6 TO 8 FEET APART IN A TRIANGLE. THEY SHOULD BE ALIVE AND AT LEAST 6" THICK. NAIL 2X6 BEAMS AROUND THE OUTSIDE OF THE TREES TO HOLD THE FLOOR AND THE ROOF. MAKE SURE THE FLOOR IS LEVEL. THE ROOF SHOULD SLANT BACK DOWN A FEW INCHES TO ALLOW THE RAIN TO RUN OFF.

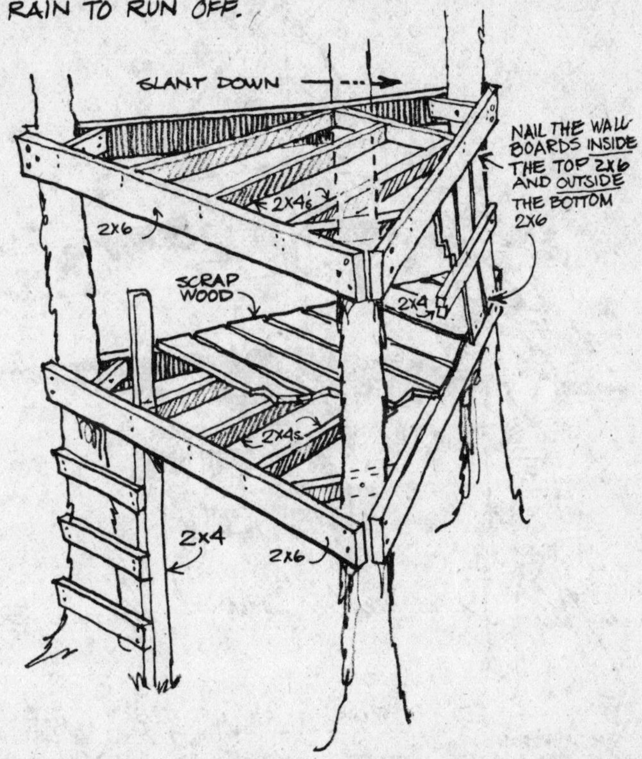

- USE 2X4s TO HELP SUPPORT THE FLOOR AND THE ROOF.
- USE ¾" THICK SCRAP WOOD TO COVER THE FLOOR AND ROOF.
- NAIL THE WALLBOARDS TO THE INSIDE OF THE TOP 2X6 AND THE OUTSIDE OF THE BOTTOM 2X6 BEAMS.
- USE 2X4s TO MAKE THE RAILINGS AND THE LADDER,
- USE ONLY GALVANIZED COATED NAILS.
- COVER THE ROOF WITH ASPHALT SHINGLES.

FORT OF THE FUTURE

THIS FORT OF THE FUTURE REQUIRES SOME SKILL TO BUILD AND A LITTLE BIT OF MONEY FOR THE MATERIALS SO YOU MAY NEED THE HELP OF AN ADULT.

SOME OF THE LUMBER (2X4s) MAY BE FOUND AT CONSTRUCTION SITES WHERE NEW HOUSES ARE BEING BUILT, BUT BE SURE TO GET THE CONTRACTOR'S PERMISSION BEFORE YOU TAKE ANY WOOD. THE BASE IS A CARDBOARD INDUSTRIAL PACKING DRUM AND CAN BE FOUND AT YOUR LOCAL MANUFACTURING PLANTS. THE WHOLE FORT IS COVERED WITH SILVER MYLAR PLASTIC FILM TO GIVE IT AN OUTER-SPACE LOOK.

MATERIALS

FOUR	4x4	6' LONG	POSTS
FOUR	2x4	8' 6" LONG	CROSSBEAMS
FOUR	2x4	6' LONG	FLOOR PERIMETER FRAME
FOUR	2x4	5' LONG	DIAGONAL UPRIGHTS
FOUR	2x4	4'6" LONG	LOWER WINDOW FRAME
FOUR	2x4	28" LONG	UPPER WINDOW FRAME
FOUR	1x3	4' 6" LONG	WINDOW STIFFENERS
FOUR	2x3	1' LONG	WINDOW PROPS
ONE	$\frac{1}{2}$"	PLYWOOD 4'x8' *	FLOOR
FOUR	$\frac{1}{2}$"	CDX PLYWOOD 4'x8' *	SIDES
ONE	32" high 20" DIAMETER PAPER PACKING DRUM		- BASE
FOUR	SETS 2" BUTT HINGES		- WINDOW HINGES
ONE	ROLL REFLECTIVE MYLAR		30' x 5' - COVERING

* THIS IS WHAT YOU ASK FOR AT THE LUMBER YARD.

BEFORE YOU BEGIN IT IS A GOOD IDEA TO DISCUSS THIS PROJECT WITH YOUR NEIGHBORS AND GET THEIR APPROVAL. PERHAPS THEY WILL LEND A HAND OR VOLUNTEER SOME MATERIALS.

FIND A SPOT THAT IS REASONABLY FLAT AND DIG FOUR 24" DEEP HOLES 12" ON CENTER WITH A POSTHOLE DIGGER.

PLACE THE DRUM AROUND THE HOLES AND INSERT THE FOUR POSTS THROUGH THE DRUM AND INTO THE HOLES. BE SURE THE POSTS ARE TOUCHING THE INSIDE WALL OF THE DRUM AND CHECK WITH A LEVEL TO BE SURE THEY ARE VERTICAL, THEN FILL THE POSTHOLES WITH CONCRETE.

MARK A LEVEL LINE ACROSS THE TOPS OF THE POSTS, EVEN WITH THE TOP OF THE DRUM, AND MAKE A NOTCH ON BOTH SIDES OF EACH POST TO ACCEPT THE CROSSBEAMS.

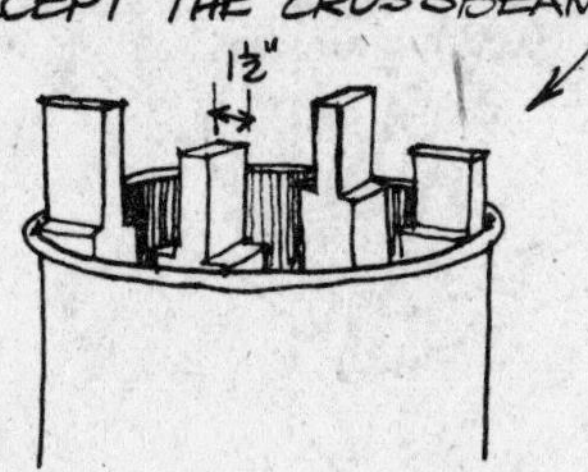

NOTCH EACH PAIR OF CROSSBEAMS SO THEY FIT TOGETHER IN THE MIDDLE.

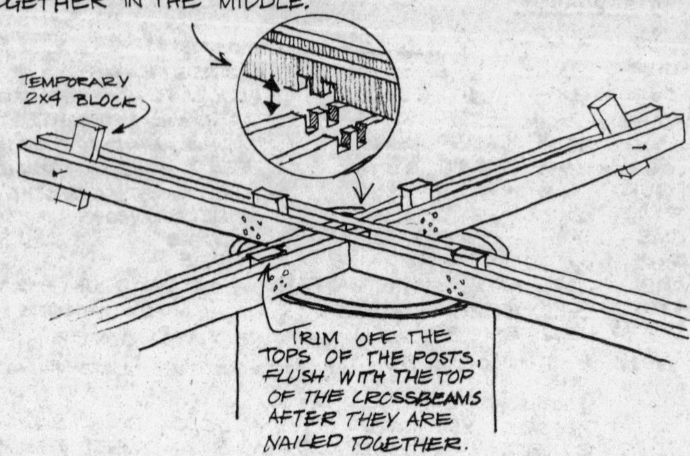

CUT & FIT 2X4s BETWEEN THE CROSSBEAMS TO FORM A SQUARE PERIMETER. THEN CUT AND NAIL FOUR DIAGONAL UPRIGHTS TO THE CENTER OF THE STRUCTURE.

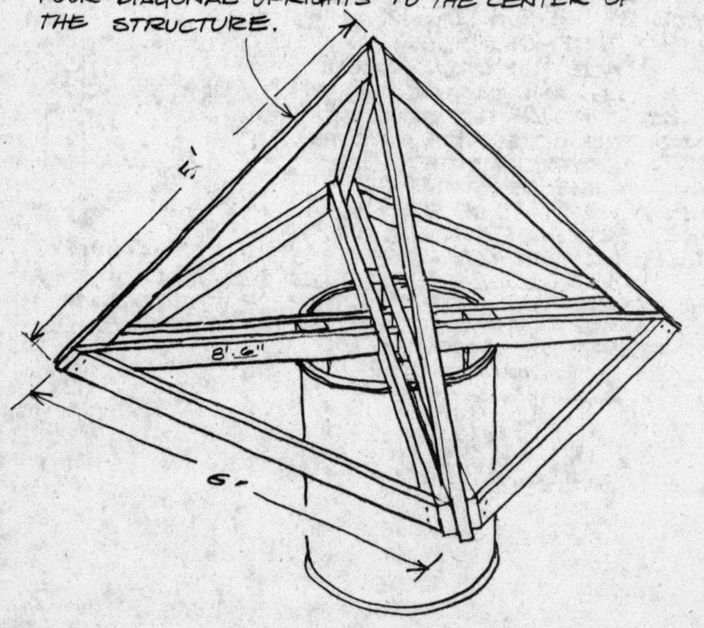

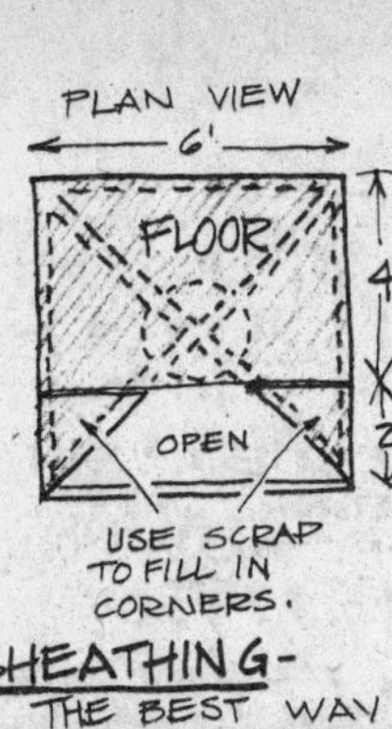

FLOOR – CUT AND NAIL A 4'X6' PANEL OF ½" PLYWOOD TO THE FRAME. USING LEFT OVER SCRAP FILL IN THE CORNERS ON EACH SIDE OF THE OPENING.

WINDOW FRAME – CUT TWO PIECES OF 2X4 FOR THE WINDOW FRAME. THE LOWER PIECE GOES IN 15" FROM THE FLOOR AND THE UPPER PIECE GOES IN 24" FROM THE TOP.

24"
15"

SHEATHING –
THE BEST WAY TO FIT THE PLYWOOD SHEATHING TO THE STRUCTURE IS TO TEMPORARILY NAIL A PANEL IN PLACE AND MARK ON THE BACK WHERE IT MEETS THE FRAME.

WINDOW PANELS –
CUT THE WINDOW PANELS SO THEY OVERLAP THE LOWER WINDOW FRAME BY ½" AND HINGE THE TOP EDGE TO THE TOP WINDOW FRAME.

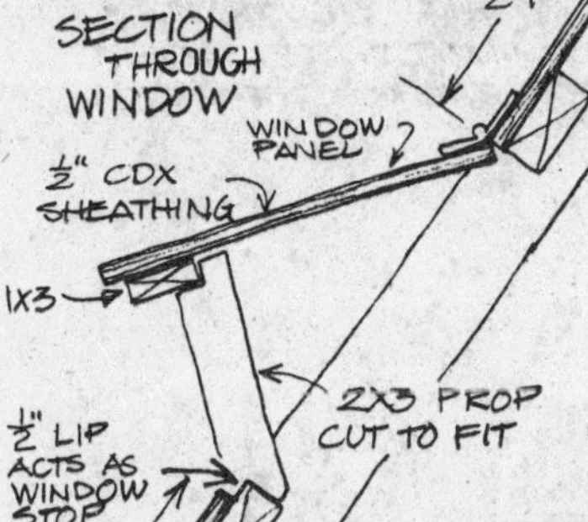

WINDOW PROP –
MAKE A PROP FROM A PIECE OF 2X3 TO HOLD THE WINDOW OPEN. NOTCH THE ENDS TO FIT THE WINDOW FRAME.

PLASTIC COVERING –
COVER THE ENTIRE STRUCTURE WITH REFLECTIVE SILVER MYLAR AND STAPLE IT TO THE CDX SHEATHING. WHEN JOINING THE PIECES TOGETHER DO IT BY CREASING THE EDGE OF BOTH PIECES AND LAPPING THEM INTO EACH OTHER.

← STAPLE THROUGH BOTH PIECES AT ONCE.

NOTE: USE NON RUSTING STAPLES.

FORT TOYS

IT MAY BECOME NECESSARY TO PROTECT YOURSELF FROM POTENTIAL "INVADERS". AN ARSENAL OF FORT TOYS CAN BE MADE WITH MATERIALS FOUND AROUND THE HOUSE, HOWEVER THEY MUST BE SAFE.

A SAFE FORT TOY DEPENDS ON THE AMMUNITION IT USES, FOR EXAMPLE, PILLOWS, INFLATABLE OBJECTS SUCH AS BALLOONS, WIFFLE BALLS, NERF-(SOFT PLASTIC FOAM) BALLS, TENNIS BALLS, PING-PONG BALLS, SNOWBALLS OR SOAP BUBBLES TO NAME A FEW. IN THE SUMMER YOU CAN USE WATER GUNS OR MAKE WATER BOMBS OUT OF BALLOONS. ALTHOUGH THESE THINGS ARE CONSIDERED SAFE THERE IS ONE RULE THAT MUST BE OBEYED BY ALL. THAT IS:

NEVER AIM OR FIRE AT THE HEAD.

MAKE IT UNDERSTOOD THAT ANY KID THAT DISOBEYS THIS RULE WILL BE IMMEDIATELY SENT HOME BY THE OTHER KIDS FOR THE REST OF THE DAY.

IN ADDITION IT SHOULD ALSO BE UNDERSTOOD THAT:

1. NO DANGEROUS GUNS SUCH AS BB GUNS ALLOWED.
2. NO MATCHES
3. NO FIRECRACKERS
4. NO TYING UP ANOTHER PERSON WITH ROPE.
5. NO PUSHING OR SHOVING OR CALLING BAD NAMES
6. NO ICE IN SNOWBALLS

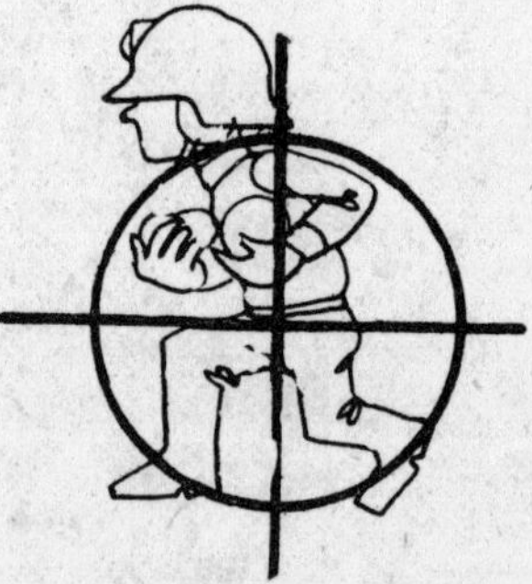

TO MAKE THIS SIMPLE CANNON YOU NEED A HEAVY CARDBOARD TUBE WHICH YOU CAN GENERALLY FIND DISCARDED BEHIND RUG STORES. THE SMALLER TUBE IS SOLD IN STATIONERY STORES FOR MAILING POSTERS, (ASK FOR A "MAILING TUBE"). THE ONLY OTHER THING YOU NEED IS A STRONG CARDBOARD BOX (LIQUOR BOX).

DRILL A HOLE (MIDWAY) THROUGH THE SIDE OF THE LARGE CARDBOARD TUBE. THEN CUT OUT THE CARDBOARD BOX AS SHOWN AND ASSEMBLE THEM TOGETHER.

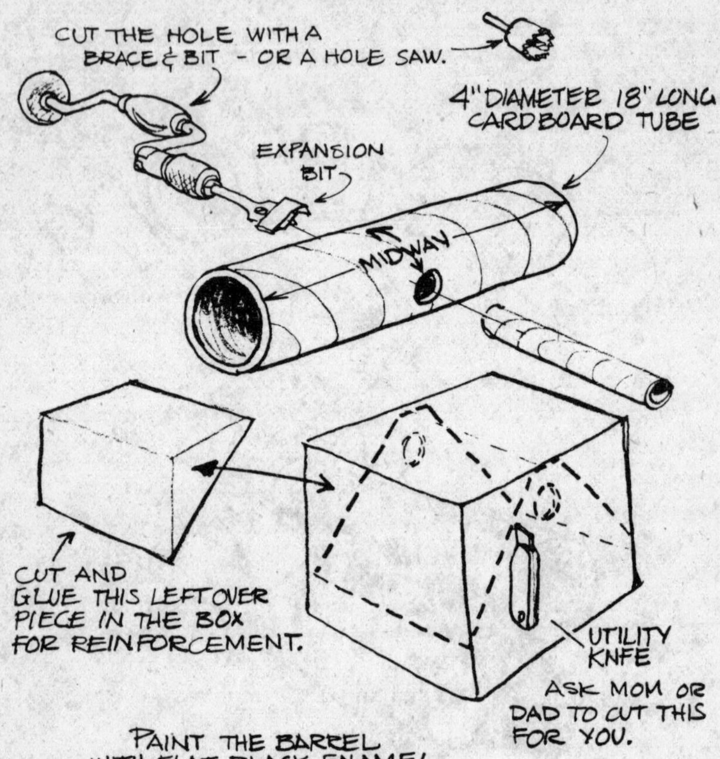

PAINT THE BARREL WITH FLAT BLACK ENAMEL AND PAINT THE BASE WITH OLIVE DRAB (GREEN & BROWN).

NOTE: ALTHOUGH PAINT WILL HELP KEEP THE CANNON WEATHERPROOF, IT IS A GOOD IDEA TO KEEP IT OUT OF THE RAIN.

BLASTER*

THIS LETHAL LOOKING WEAPON ACTUALLY DOES NOTHING BUT LOOK SCAREY. IT IS MADE FROM DISCARDED MATERIALS FOUND AT YOUR LOCAL DUMP.

BEGIN BY LOOKING FOR THE BASE OF AN OLD SWIVEL CHAIR. IF YOU CAN NOT FIND ONE, THEN MAKE A BASE OUT OF SCRAP LUMBER LIKE THIS.

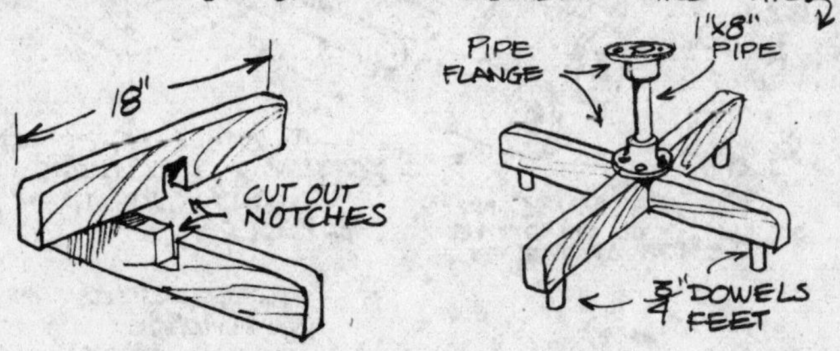

* ACCESSORY FOR THE FORT OF THE FUTURE

NEXT MAKE A CRADLE OUT OF SCRAP PLYWOOD TO HOLD THE BARREL.

12" 14" 12"

¾" PLYWOOD

BOTTOM 5½" 12½

FIND OR MAKE A BOX TO FIT INSIDE THE CRADLE.

CENTER LINE

POLE

5½ 10" 12"

DRILL A HOLE AND INSERT A POLE THROUGH THE SIDE NEAR THE TOP.

DRILL A PAIR OF HOLES IN THE FRONT AND INSERT TWO MORE POLES.

BARREL

SCREW BITS OF WOOD ON TO MAKE SIGHTS.

7" 14"

4" SOUP CANS NAIL ONTO THE ENDS OF THE POLES.

CUT HOLES OUT OF THE BOTTOM OF TWO LARGE JUICE CANS AND SLIDE THEM ONTO THE POLES.

PISTOL GRIP CUT FROM ¾" PLYWOOD

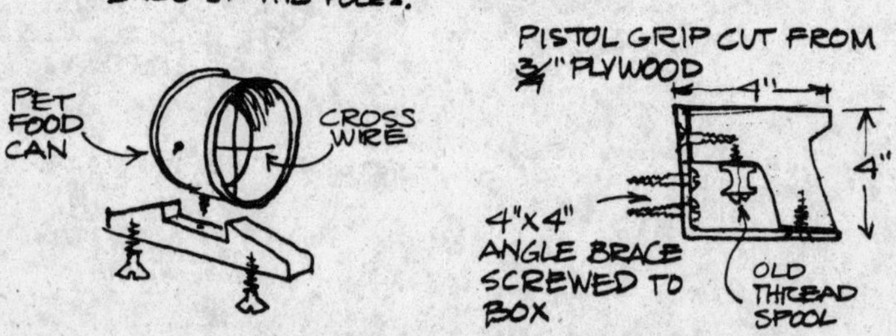

CAT-A-PULT

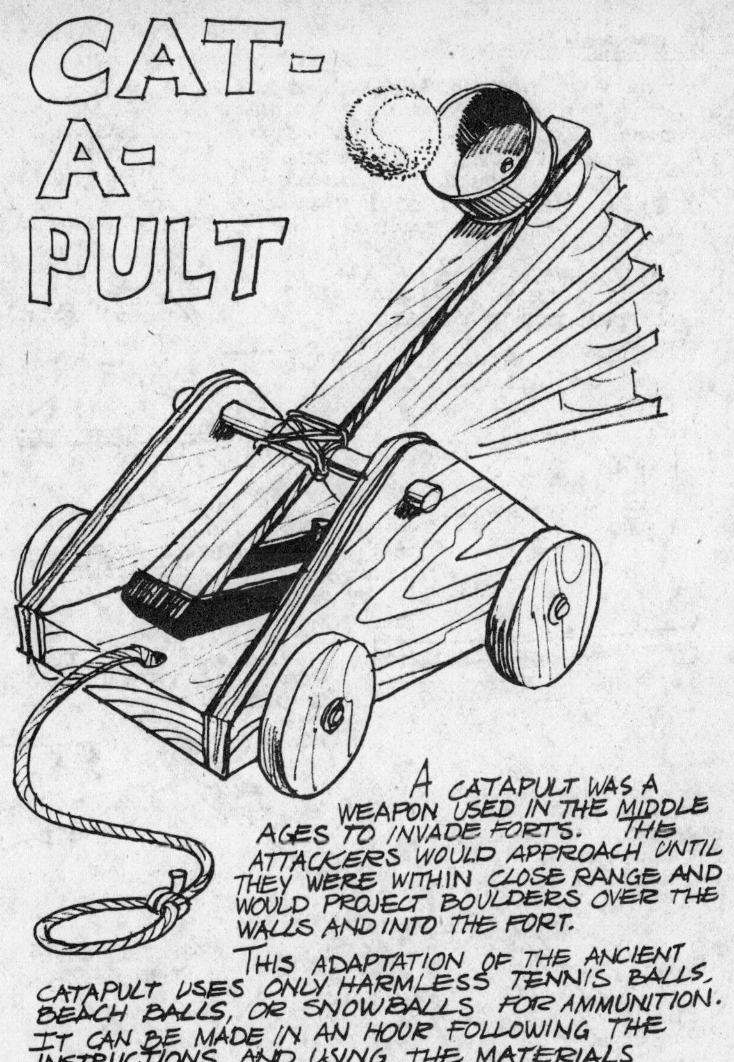

A CATAPULT WAS A WEAPON USED IN THE MIDDLE AGES TO INVADE FORTS. THE ATTACKERS WOULD APPROACH UNTIL THEY WERE WITHIN CLOSE RANGE AND WOULD PROJECT BOULDERS OVER THE WALLS AND INTO THE FORT.

THIS ADAPTATION OF THE ANCIENT CATAPULT USES ONLY HARMLESS TENNIS BALLS, BEACH BALLS, OR SNOWBALLS FOR AMMUNITION. IT CAN BE MADE IN AN HOUR FOLLOWING THE INSTRUCTIONS AND USING THE MATERIALS DESCRIBED ON THE NEXT PAGES.

MATERIALS:

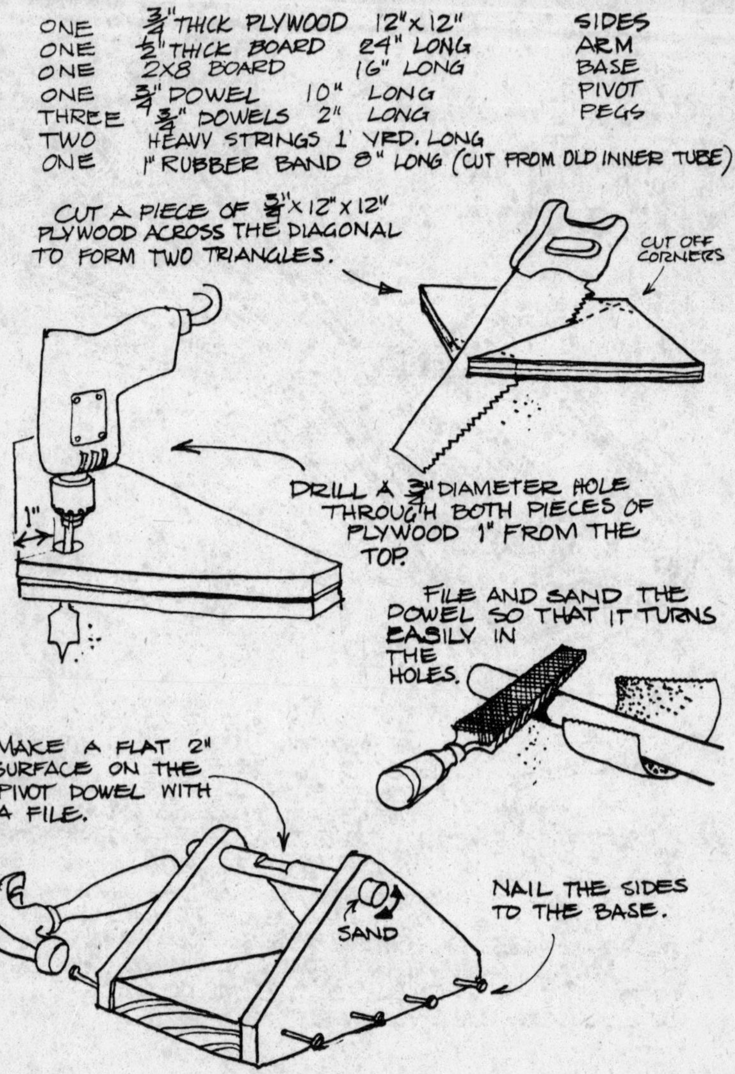
ONE 3/4" THICK PLYWOOD 12"x12" SIDES
ONE 1/2" THICK BOARD 24" LONG ARM
ONE 2x8 BOARD 16" LONG BASE
ONE 3/4" DOWEL 10" LONG PIVOT
THREE 3/4" DOWELS 2" LONG PEGS
TWO HEAVY STRINGS 1 YRD. LONG
ONE 1" RUBBER BAND 8" LONG (CUT FROM OLD INNER TUBE)
CUT A PIECE OF 3/4"x12"x12" PLYWOOD ACROSS THE DIAGONAL TO FORM TWO TRIANGLES.
CUT OFF CORNERS
1"
DRILL A 3/4" DIAMETER HOLE THROUGH BOTH PIECES OF PLYWOOD 1" FROM THE TOP.
FILE AND SAND THE DOWEL SO THAT IT TURNS EASILY IN THE HOLES.
MAKE A FLAT 2" SURFACE ON THE PIVOT DOWEL WITH A FILE.
SAND
NAIL THE SIDES TO THE BASE.

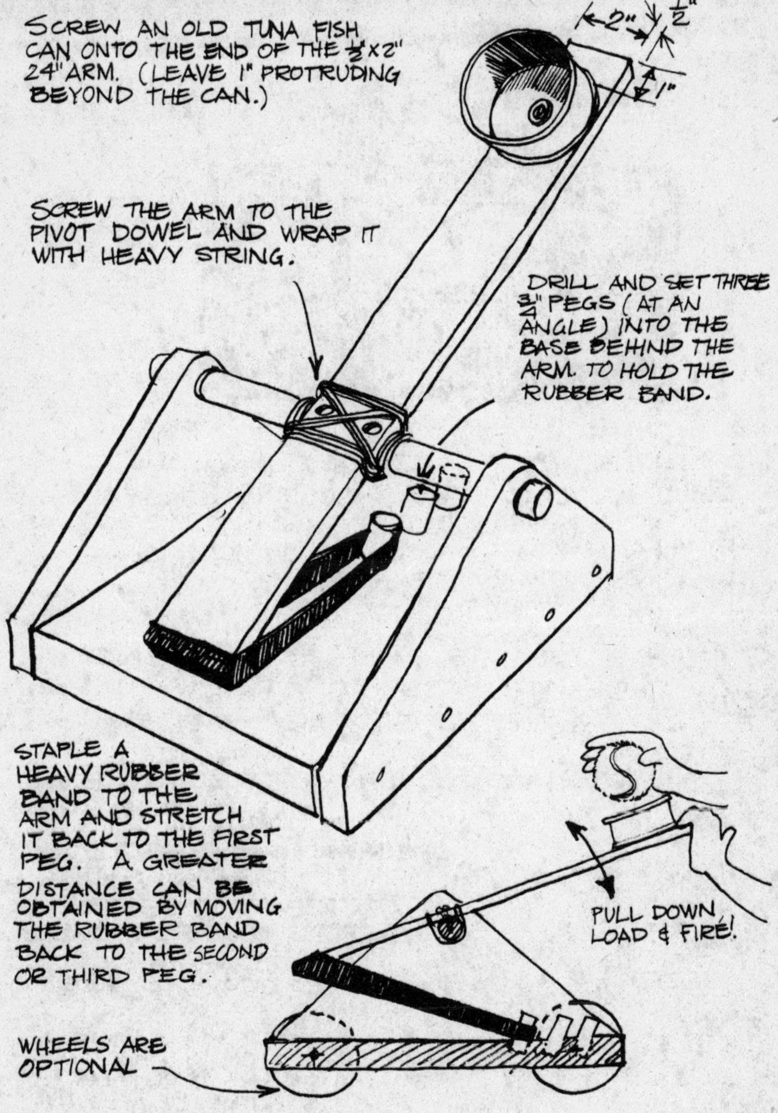
SCREW AN OLD TUNA FISH CAN ONTO THE END OF THE ½"x2" 24" ARM. (LEAVE 1" PROTRUDING BEYOND THE CAN.)
2"
½"
1"
SCREW THE ARM TO THE PIVOT DOWEL AND WRAP IT WITH HEAVY STRING.
DRILL AND SET THREE ¾" PEGS (AT AN ANGLE) INTO THE BASE BEHIND THE ARM. TO HOLD THE RUBBER BAND.
STAPLE A HEAVY RUBBER BAND TO THE ARM AND STRETCH IT BACK TO THE FIRST PEG. A GREATER DISTANCE CAN BE OBTAINED BY MOVING THE RUBBER BAND BACK TO THE SECOND OR THIRD PEG.
PULL DOWN, LOAD & FIRE!
WHEELS ARE OPTIONAL

TANKS ARE FUN TO GET INTO AND CRAWL AROUND THE HOUSE. ONCE YOU ARE INSIDE NO ONE CAN SEE YOU AND YOU CAN MAKE IT MOVE BY WALKING ON YOUR KNEES.

THIS TANK IS EASY TO MAKE SINCE ALL THE MATERIALS ARE THROW-OUT THINGS FROM THE GROCERY STORE.

MATERIALS:
- 1 LARGE CARDBOARD BOX BIG ENOUGH TO CRAWL INTO
- 1 SMALL CARDBOARD BOX (SQUARE SHAPE)
- 1 CARDBOARD TUBE FROM A ROLL OF PAPER TOWELS

TOOLS:
- 1 BOTTLE OF WHITE GLUE
- 1 SMALL KITCHEN KNIFE

PAINT:
- 1 SMALL CAN OF BLACK
- 1 SMALL CAN OF GREEN
- 1 SMALL CAN OF BROWN

1. SEAL THE OPEN END OF THE BIG BOX WITH GLUE AND ASK MOM OR DAD TO CUT A HOLE LARGE ENOUGH FOR YOUR KNEES AND LEGS TO FIT THROUGH.

2. TURN THE BOX OVER AND CUT A HOLE BIG ENOUGH FOR YOUR HEAD TO FIT THROUGH.

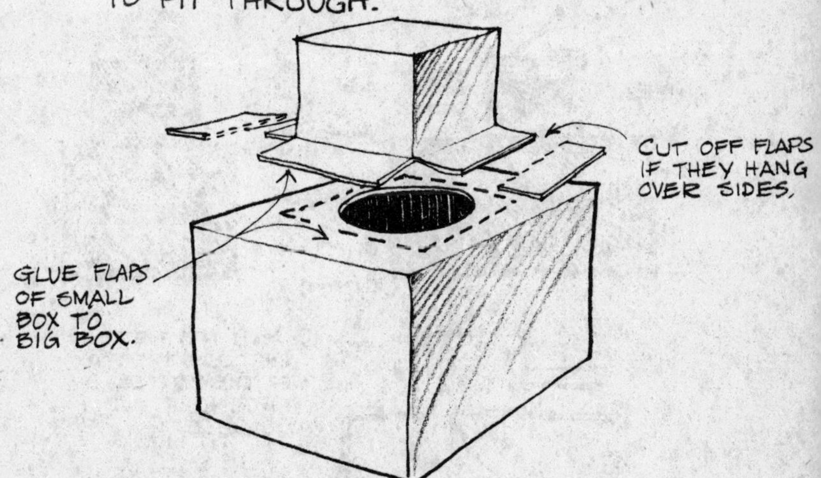

GLUE THE SMALL BOX OVER THE HOLE OF THE BIG BOX.

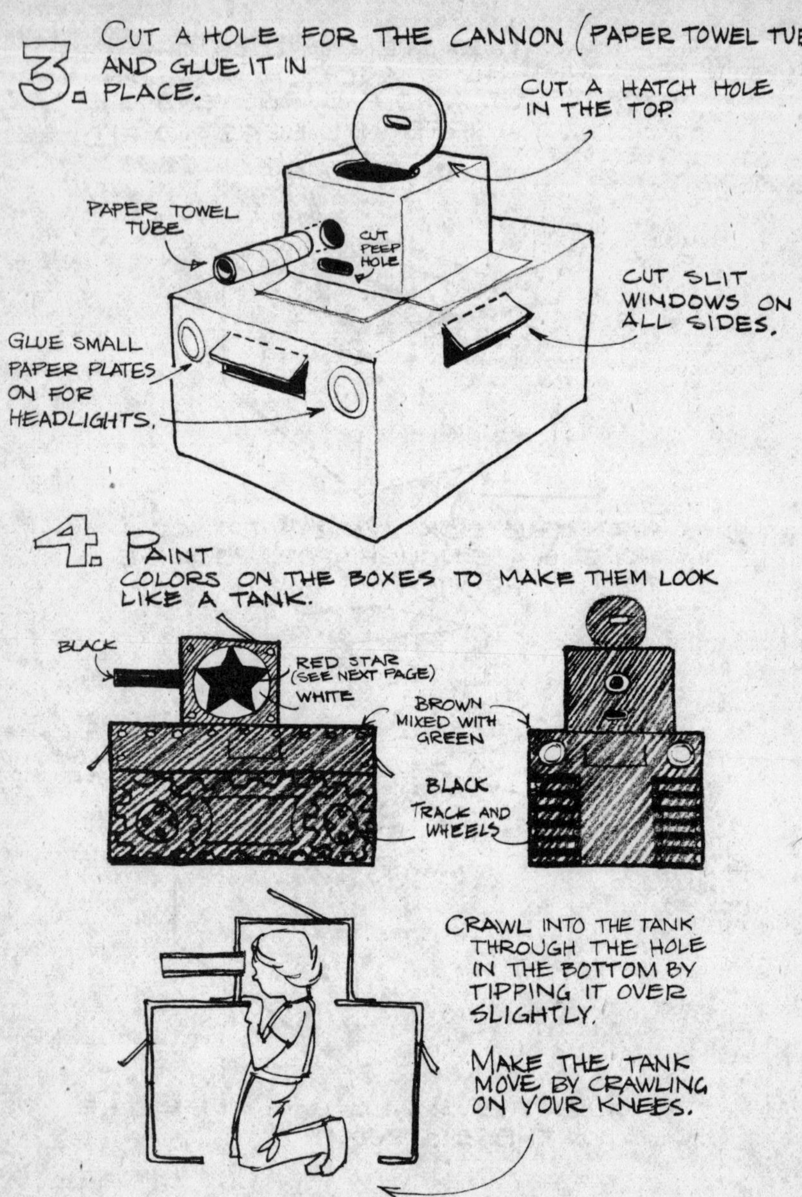
3. CUT A HOLE FOR THE CANNON (PAPER TOWEL TUBE) AND GLUE IT IN PLACE.
CUT A HATCH HOLE IN THE TOP.
PAPER TOWEL TUBE
CUT PEEP HOLE
CUT SLIT WINDOWS ON ALL SIDES.
GLUE SMALL PAPER PLATES ON FOR HEADLIGHTS.
4. PAINT COLORS ON THE BOXES TO MAKE THEM LOOK LIKE A TANK.
BLACK
RED STAR (SEE NEXT PAGE)
WHITE
BROWN MIXED WITH GREEN
BLACK
TRACK AND WHEELS
CRAWL INTO THE TANK THROUGH THE HOLE IN THE BOTTOM BY TIPPING IT OVER SLIGHTLY.
MAKE THE TANK MOVE BY CRAWLING ON YOUR KNEES.

HOW TO MAKE A FIVE-POINTED STAR

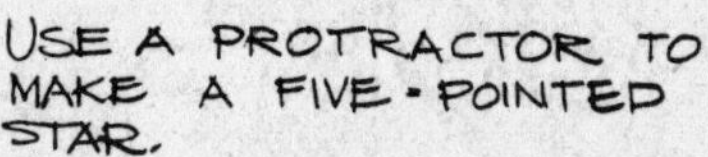

USE A PROTRACTOR TO MAKE A FIVE-POINTED STAR.

SINCE THERE ARE 360° (DEGREES) IN A COMPLETE CIRCLE DIVIDE BY 5 (POINTS)

$$5\overline{)360}\ = 72^\circ$$

- LAY THE PROTRACTOR ON A CLEAN, WHITE PIECE OF PAPER. DRAW AROUND IT WITH A PENCIL.
- FLIP IT OVER AND TRACE THE OTHER HALF OF THE CIRCLE.
- PUT A MARK ANYWHERE ON THE CIRCLE AND MARK OFF 72° USING THE PROTRACTOR.
- DO THIS FIVE TIMES.
- USING A RULER, CONNECT ALL THE POINTS AND COLOR THE STAR RED WITH A MAGIC MARKER.

NOTE: TO MAKE IT EVEN BETTER USE WHITE & RED CONTACT PAPER INSTEAD.

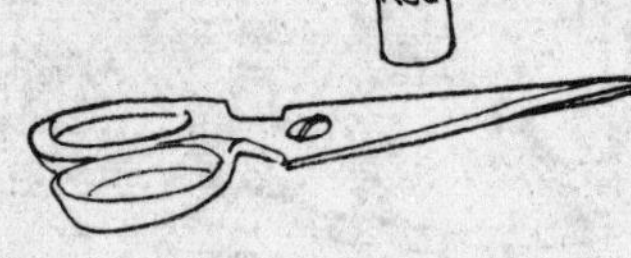

FORT MODEL

... MADE FROM PAPIER-MÂCHÉ OVER CHICKEN WIRE

MODEL FORTS

ARE FUN TO MAKE AND FUN TO PLAY WITH WHEN THEY ARE FINISHED. YOU CAN BUY MINATURE TOY KNIGHTS AND HORSES AND PUT THEM IN THE FORT. TOY STORES ALSO CARRY AN ARRAY OF SPECIAL SCALED-DOWN MEDIEVAL WAR TOYS LIKE CROSSBOWS, CATAPULTS, MORTARS, CANNONS AND LADDERS FOR THE INVADING FORCES TO SCALE THE WALLS.

START WITH A PIECE OF ½" PLYWOOD 24"X24" AS A BASE AND LAY OUT YOUR FORT WITH A PENCIL. MARK WHERE YOUR GATEHOUSE (ENTRANCE), GREAT HALL, STABLE, STOREHOUSE, CHAPEL AND CORNER TOWERS SHOULD BE. A REAL FORT GENERALLY HAD AN OUTER AND AN INNER WALL OF DEFENSE, HOWEVER YOU CAN HAVE ONLY ONE WALL IN ORDER TO KEEP IT SIMPLE.

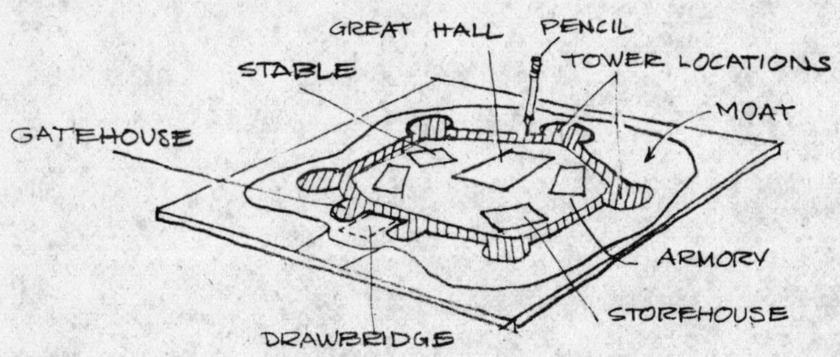

BUILD THE WALLS AND TOWERS OUT OF 1" DIAMETER CHICKEN WIRE STAPLED TO THE PLYWOOD BASE AND COVER IT WITH PAPIER-MÂCHÉ.

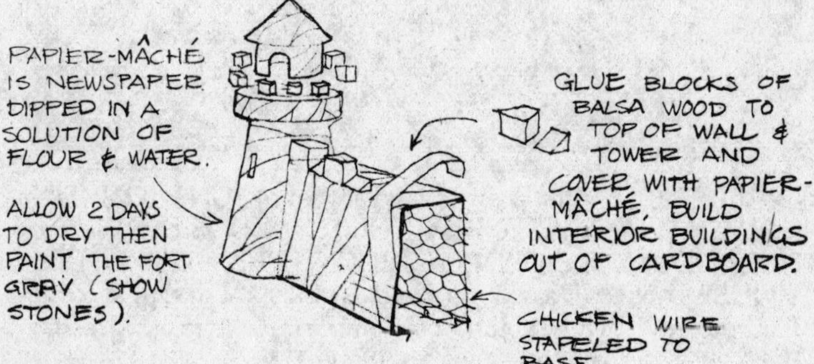

DAVID STILES is the designer and author of FUN PROJECTS FOR DAD AND THE KIDS, HUTS AND HIDEAWAYS, EASY-TO-MAKE CHILDREN'S FURNITURE, and THE TREE HOUSE BOOK. He is a graduate of industrial design at Pratt Institute, and studied at the Accademi Di Belle Arte in Florence, Italy. Mr. Stiles lives in New York City.